Air Fryer

Cookbook

For Beginners

Amazingly Easy And Healthy Recipes To Roast, Grill, Bake And Fry For You And Your Family. How To Cook Crispy Fries, Vegetables, Fish, Meat With Much Less Oil And Get The Best Results In A Short Time

Ivy Cook

Table of contents

By reading this document, the reader agrees that under no circumstances is the author responsible for any losses, direct or indirect, which are incurred as a result of the use of information contained within this document, including, but not limited to, — errors, omissions, or inaccuracies.

Introduction

You've made the best first step to reducing your meal preparation time by purchasing the Air Fryer Cookbook. You can locate your favorite recipes and whip up a remarkable meal at home in half of the time. All you need to do is 'punch in' the temperature and times. That is only the beginning of your journey with your new Air Fryer recipes. You'll be glad you are beginning a new way of cooking:

It won't be necessary to add oil to the cooker if you have frozen products that are meant for baking. You only need to adjust the timer and cook. All of the excess fat will drip away into a tray beneath the basket.

You can cook whatever meat you enjoy and receive delicious and healthy results. You will understand this once you begin trying out some of these new recipes.

For example, you can cook French fries with a tablespoon of oil versus a vat of oil.

You only need to remove the cooking bowl, drip pan, or the cooking basket. It is inside a cover, which means you won't have oil vapor deposits on the walls, floors, or countertops.

You can use the dishwasher to clean the movable parts. You can also use a sponge to clean the bits of food that might be stuck to the AF surfaces.

It is possible to splurge on the more expensive oils since you only use such a minimal amount.

The Air Fryer is capable of functioning like so many products, whether you need an oven, a hot grill, a toaster, a skillet, or a deep fryer—it is your answer!

The machine will automatically shut down when the cooking time is completed. You will have less burned or overheated food items. The fryer will not slip because of the non-slip feet, which help eliminate the risk of the machine from falling off of the countertop. The closed cooking system helps prevent burns from hot oil or other foods. Air Fryer heats and cooks foods by circulating extremely high-temperature heat, up to 400 degrees F, at high-speed. At such a high-speed circulation, it only consumes a negligible amount of oil, usually about one tablespoon to prepare aromatic crispy foods

The hot-air technology cooks foods from different angles. More importantly, the technology maintains their great taste and essential nutrients. Meals prepared with such a meager amount of oil also free you from greasy stains on your fingers. This technology has brought a new era of cooking by using 80% less fat as compared to traditional deep fat frying. Air Fryer comes with an exhaust fan placed right above the cooking chamber; this fan provides the food required airflow. This modern heating technology ensures that food is cooked with constant heated air

As a result, the same heating temperature is maintained covering different parts of the added food ingredients. Air fryer is completely odorless and harmless making it a user-friendly kitchen invention.

It is a complete guide for your Air Fryer Cooking!

How to get out the most of your air fryer

•Firstly, you are able to set the temperature based on the thing you need and allow it to preheat for just a few minutes before you decide to place the food in.

•Then open it up and put the meals within the air fryer. Close it and allow the food to prepare based on the timing you've set. It will not take enough time regardless of what you're cooking.

•You may use a little bit of cooking spray or splash some oil around the food before putting it in. This can help to prevent the meals from getting stuck towards the pan. The oil likewise helps to create food a bit crispier and provides the taste of standard fried food.

•Halfway with the set time or perhaps in between times, provide the air fryer just a little shake so the air inside circulates easily and your meals are completely cooked.

•Other than frying, you can test other cooking methods while using additional parts the air flyers usually include. You should use the grill or baking tray in compliance using the instructions that include the environment fryer.

•Another factor to bear in mind is you shouldn't overcrowd the fryer by putting an excessive amount of food inside it. Put small batches of food within the air fryer to ensure that there's enough room for this all to maneuver and prepare evenly. Overcrowding won't let it move and a few parts is going to be left uncooked. The cooking can also be elevated if you devote an excessive amount of at the same time.

•In the situation of marinated food, allow it to be as dry as you possibly can before placing it within the air fryer. Wet food may cause splattering in addition to an excessive amount of smoke being released in the fryer.

•The separator that will get the environment fryer can be put among layers of food in order to prepare various things concurrently. Just make certain the temperature needed is identical for those products or they will not be cooked evenly.

•Pre-packaged meals may also be made while using air fryer. Lower the suggested oven temperature and hang the environment fryer after your meals are placed within it. Additionally, it cuts down on the normal cooking considerably.

•For baking food, you should use the baking pan provided or purchase one individually. You do not need a stove to bake some muffins any longer and also you have them cooked even faster than in the past.

•Roasting meals are also simpler than using conventional methods. It doesn't take considerable time but you just get individuals healthy and attractive roasted veggies or meats that you simply love. For grilling food, put the grill layer inside as well as your food on the top from it. You don't need to help keep flipping the meals over as with a conventional grill.

This will make things much easier. Give it a shake following a couple of moments for much better air flow.

As possible clearly see in the information above, significantly less efforts are needed while the food is cooking. Make use of the simple mechanism and prepare a number of different meals every single day.

The benefits of air fryer

Air fryer has several benefits that make this kitchen appliance incredibly effective and literally master-have appliance

Low Fat Cooking

Perhaps, the most valuable advantage of air fryer is that it provides you with a finished product that is very similar to the usual fried food, with minimal fat. Due to the classic frying methods a large amount of oil is used, which provides crispy crust, but at the same time - more harmful fat and extra calories.

With the air fryer you can cook your favorite fried meals, such as French fries or fried chicken wings with the same crispy crust and low fat.

Safety cooking

Unlike traditional frying, where hot oil sprays in all directions and you have plenty of ways to burn yourself, cooking in air fryer is completely safe. This device does not have outdoor areas where you can accidentally get burned.

Quickly warms up

A distinctive feature of air fryer is the rapid heating of the device. Usually it takes only 2-3 minutes to reach the required temperature and start cooking.

Variety of cooking

Thanks to air fryer you can cook a huge amount of dishes depending on your wishes. For example, you can cook ribs, vegetable dishes, beef, pork or lamb. Moreover, you can prepare a variety of desserts, such as fruit chips, pies, fruit bread and many other delicious and healthy food. By the way, air fryer can easily replace some traditional kitchen appliances and become indispensable in your kitchen.

Easy to clean

Most kitchen appliances require a lot of effort to clean them after cooking. But not air fryer. You won't have any problems with this appliance. Not only cook quickly but it also cleans quickly. Once it has cooled down, you can disassemble the air fryer for spare parts and put them in the dishwasher. Also, due to the minimal use of oil during cooking, the parts of the oven are not very dirty, which allows you to wash the device in minutes.

Tips on cleaning

The first thing you should have at your fingertips is that, if you do not clean and maintain your air fryer from time to time, it won't last long. Following these guidelines will secure the fact that your air fryer will remain effective and durable for years to come

How to clean your air fryer:

1.   Unplug your air fryer from the wall socket and allow it to cool until you can touch.
2.   Using a wet rag, wipe the exterior part of your air fryer
3.   Remove the air fryer pan, tray, basket and wash it with hot water and a dishwasher soap in your sink. These parts are removable and are safe for an easy cleanup.
4.   Use a cloth or sponge to wipe and clean the inner part of your air fryer
5.   If you find any ingredients sticking in your air fryer, scrub it off with a brush.
6.   Before adding the pan, tray, and basket back into your air fryer ensure they are entirely dry
7.   Once your air fryer is cleaned, store it safely.

How to maintain your air fryer:

Your air fryer requires a standard form of maintenance to ensure it does not get damaged or work erroneously. To do this, one needs to follow this instruction

1.   Before using your air fryer, make sure you check the cord. That is, do not plug a damaged cord into an outlet; this can result in a ghastly injury or even death.
2.   Make sure your air fryer is clean and free of any debris before you begin cooking. Check the inner part and make sure you remove anything redundant in there.
3.   Ensure the air fryer is placed upright, on a flat surface.
4.   Make sure that your air fryer is not too close to the wall or another appliance. Air fryers require 4-inches of space all around them.
5.   One after the other, check each component of your air fryer, including the basket, pan, and handle.

BREAKFAST RECIPES

Crunchy Blueberry Bread Bites

Preparation Time: 18 Minutes

Servings: 5

Ingredients:

Milk-1/3 cup

Cream cheese; whipped-4 tbsp

Nutmeg; ground-1/4 tsp

Blueberries -1/4 cup

Corn flakes; crumbled-1 ½ cups

Sugar-3 tsp

Eggs; whisked -2

Bread slices -5

Directions:

Whisk eggs with milk, nutmeg, and sugar in a bowl.

Mix cream cheese with blueberries in a bowl.

Spread corn flakes in a separate bowl.

Top the bread slices with cream cheese mixture then dip them in the eggs.

Coat the slices with the corn flakes and place them in the air fryer basket,

Seal the fryer, and Air fry them for 8 minutes at 400 0F.

Enjoy.

Nutrition:

Calories: 300; Fat: 5g; Fiber: 7g; Carbs: 16g; Protein: 4g

Broccoli Egg Casserole

Preparation Time: 30 Minutes

Servings: 2

Ingredients:

Broccoli head; florets separated and steamed-1

Tomato; chopped-1

Thyme; chopped -1 tsp

Carrots; chopped and steamed-3

Cheddar cheese; grated-2 oz.

Milk-2 oz.

Parsley; chopped-1 tsp

Eggs-2

Salt and black pepper - to the taste

Directions:

Whisk eggs with parsley, milk, thyme, pepper, and salt in a bowl.

Toss in the broccoli, tomato, and carrots.

Spread these vegetables in the air fryer basket.

Top them with cheddar cheese and seal the air fryer.

Cook them for 20 minutes at 350 oF.

Enjoy.

Nutrition:

Calories: 214; Fat: 4g; Fiber: 7g; Carbs: 12g; Protein: 3g

Pine Nut Spinach Pockets

Preparation Time: 14 Minutes

Servings: 2

Ingredients:

Baby spinach leaves; roughly chopped-1 lb.

Filo pastry-4 sheets

Ricotta cheese-1/2 lb.

Pine nuts -2 tbsp

Egg; whisked-1

Zest from 1 lemon; grated

Greek yogurt for serving

Salt and black pepper - to the taste

Directions:

Mix cheese, lemon zest, egg, salt, pepper, pine nuts, and spinach in a bowl.

Spread the filo sheets on the working surface and divide the spinach mixture over them.

Fold each sheet diagonally and seal the parcels.

Place these parcels in the air fryer basket and seal the fryer.

Cook them for 4 minutes at the air fryer mode at 400 0F.

Serve.

Nutrition:

Calories: 182; Fat: 4g; Fiber: 8g; Carbs: 9g; Protein: 5g

Jam Glazed French Toast

Preparation Time: 30 Minutes

Servings: 6

Ingredients:

Blackberry jam; warm -1 cup

Half and half cream-2 cups

Brown sugar-1/2 cup

Vanilla extract-1 tsp

Bread loaf; cubed-12 oz.

Cream cheese; cubed-8 oz.

Eggs-4

Cinnamon powder-1 tsp

Cooking spray

Directions:

Add cooking oil to the air fryer pan and preheat it at 300 0F.

Spread the blueberry jam at the bottom of the air fryer.

Top this jam with half of the bread cubes.

Add cream cheese on top and then add the remaining bread cubes.

Whisk eggs with half and half, vanilla, cinnamon, and sugar in a bowl.

Pour this over the bread and seal the air fryer.

Cook them for 20 minutes at 350 0F on Air fryer mode.

Enjoy.

Nutrition:

Calories: 215; Fat: 6g; Fiber: 9g; Carbs: 16g; Protein: 6g

Chive and Mushroom Omelet

Preparation Time: 25 Minutes

Servings: 4

Ingredients:

Egg whites-1 cup

Mushrooms; chopped-1/4 cup

Chives; chopped-2 tbsp

1/4 cup tomato; chopped

Skim milk-2 tbsp

Salt and black pepper - to the taste

Directions:

Whisk egg whites with milk, tomato, chives, mushrooms, pepper and salt in a bowl.

Place this mixture in the Air fryer pan and seal the fryer.

Cook the mushrooms for 15 minutes at 320 0F on-air fryer mode.

Enjoy.

Nutrition:

Calories: 100; Fat: 3g; Fiber: 6g; Carbs: 7g; Carbs: 4g

Gouda Tomato Quiche

Preparation Time: 40 Minutes

Servings: 1

Ingredients:

Yellow onion; chopped-2 tbsp

Gouda cheese; shredded -1/2 cup

Tomatoes; chopped-1/4 cup

Eggs-2

Milk-1/4 cup

Salt and black pepper - to the taste

Cooking spray

Directions:

Take a ramekin and grease it with cooking spray.

Add eggs, cheese, milk, onion, tomatoes, salt, and pepper.

Stir well until mixed then place it in the air fryer.

Cook the eggs for 30 minutes at 340 0F on Air fryer mode.

Enjoy.

Nutrition:

Calories: 241; Fat: 6g; Fiber: 8g; Carbs: 14g; Protein: 6g

Breakfast Cream And Egg Soufflé

Preparation Time: 18 Minutes

Servings: 4

Ingredients:

Eggs; whisked -4

Heavy cream-4 tbsp

Parsley; chopped-2 tbsp

Chives; chopped-2 tbsp

A pinch of red chili pepper; crushed

Salt and black pepper - to the taste

Directions:

Whisk eggs with cream, red chili pepper, salt, pepper, chives and parsley in a bowl.

Mix well then divide the mixture into 4 souffle dishes.

Place these dishes in the air fryer then secure the fryer.

Cook them for 8 minutes at 350 0F on Air fryer mode.

Enjoy.

Nutrition:

Calories: 300; Fat: 7g; Fiber: 9g; Carbs: 15g; Protein: 6g

Preparation Time: 50 Minutes

Servings: 6

Ingredients:

Milk -1 cup

Sugar-1/4 cup

Egg -1

Butter-4 tbsp

Flour-3 ¼ cups

Yeast -2 tsp

For the filling:

Cream cheese; soft-8 oz.

Raspberries-12 oz.

Vanilla extract-1 tsp

Sugar-5 tbsp

Cornstarch-1 tbsp

Zest from 1 lemon; grated

Directions:

Mix flour with yeast and sugar in a bowl.

Add egg and milk then stir well until it forms a dough.

Leave in the bowl for 30 minutes then knead again.

Whisk cream cheese with vanilla, lemon zest, and sugar.

Spread the dough on the working surface and top it with cream cheese mixture.

Roll the dough over the cream cheese filling.

Slice the roll into thick pieces and place them in the air fryer basket.

Spray them with cooking oil and seal the fryer.

Cook them for 30 minutes at 350 0F on Air fryer mode.

Enjoy.

Nutrition:

Calories: 261; Fat: 5g; Fiber: 8g; Carbs: 9g; Protein: 6g

SNACK & APPETIZERS

Buffalo Cauliflower Wings

Preparation Time: 10 minutes

Cooking Time: 14 minutes

Serve:4

Ingredients:

1 cauliflower head, cut into florets

1 tbsp butter, melted

1/2 cup buffalo sauce

Pepper

Salt

Directions:

Spray air fryer basket with cooking spray.

In a bowl, mix together buffalo sauce, butter, pepper, and salt.

Add cauliflower florets into the air fryer basket and cook at 400 F for 7 minutes.

Transfer cauliflower florets into the buffalo sauce mixture and toss well.

Again, add cauliflower florets into the air fryer basket and cook for 7 minutes more at 400 F.

Serve and enjoy.

Nutrition:

Calories 44

Fat 3 g

Carbohydrates 3.8 g

Sugar 1.6 g

Protein 1.3 g

Cholesterol 8 mg

Crunchy Bacon Bites

Preparation Time: 5 minutes

Cooking Time: 10 minutes

Serve:4

Ingredients:

4 bacon strips, cut into small pieces

1/2 cup pork rinds, crushed

1/4 cup hot sauce

Directions:

Add bacon pieces in a bowl.

Add hot sauce and toss well.

Add crushed pork rinds and toss until bacon pieces are well coated.

Transfer bacon pieces in air fryer basket and cook at 350 F for 10 minutes.

Serve and enjoy.

Nutrition:

Calories 112

Fat 9.7 g

Carbohydrates 0.3 g

Sugar 0.2 g

Protein 5.2 g

Cholesterol 3 mg

Perfect Crab Dip

Preparation Time: 5 minutes

Cooking Time: 7 minutes

Serve:4

Ingredients:

1 cup crabmeat

2 tbsp parsley, chopped

2 tbsp fresh lemon juice

2 tbsp hot sauce

1/2 cup green onion, sliced

2 cups cheese, grated

1/4 cup mayonnaise

1/4 tsp pepper

1/2 tsp salt

Directions:

In a 6-inch dish, mix together crabmeat, hot sauce, cheese, mayo, pepper, and salt.

Place dish in air fryer basket and cook dip at 400 F for 7 minutes.

Remove dish from air fryer.

Drizzle dip with lemon juice and garnish with parsley.

Serve and enjoy.

Nutrition:

Calories 313

Fat 23.9 g

Carbohydrates 8.8 g

Sugar 3.1 g

Protein 16.2 g

Cholesterol 67 mg

Herb Zucchini Slices

Preparation Time: 10 minutes

Cooking Time: 15 minutes

Servings: 4

Ingredients:

2 zucchinis, slice in half lengthwise and cut each half through middle

1 tbsp olive oil

4 tbsp parmesan cheese, grated

2 tbsp almond flour

1 tbsp parsley, chopped

Pepper

Salt

Directions:

Preheat the air fryer to 350 F.

In a bowl, mix together cheese, parsley, oil, almond flour, pepper, and salt.

Top zucchini pieces with cheese mixture and place in the air fryer basket.

Cook zucchini for 15 minutes at 350 F.

Serve and enjoy.

Nutrition:

Calories 157

Fat 11.4 g

Carbohydrates 5.1 g

Sugar 1.7 g

Protein 11 g

Cholesterol 20 mg

Curried Sweet Potato Fries

Preparation Time: 10 minutes

Cooking Time: 20 minutes

Servings: 3

Ingredients:

2 small sweet potatoes, peel and cut into fries shape

1/4 tsp coriander

1/2 tsp curry powder

2 tbsp olive oil

1/4 tsp sea salt

Directions:

Add all ingredients into the large mixing bowl and toss well.

Spray air fryer basket with cooking spray.

Transfer sweet potato fries in the air fryer basket.

Cook for 20 minutes at 370 F. Shake halfway through.

Serve and enjoy.

Nutrition:

Calories 118

Fat 9 g

Carbohydrates 9 g

Sugar 2 g

Protein 1 g

Cholesterol 0 mg

Roasted Almonds

Preparation Time: 5 minutes

Cooking Time: 8 minutes

Servings: 8

Ingredients:

2 cups almonds

1/4 tsp pepper

1 tsp paprika

1 tbsp garlic powder

1 tbsp soy sauce

Directions:

Add pepper, paprika, garlic powder, and soy sauce in a bowl and stir well.

Add almonds and stir to coat.

Spray air fryer basket with cooking spray.

Add almonds in air fryer basket and cook for 6-8 minutes at 320 F. Shake basket after every 2 minutes.

Serve and enjoy.

Nutrition:

Calories 143

Fat 11.9 g

Carbohydrates 6.2 g

Sugar 1.3 g

Protein 5.4 g

Cholesterol 0 mg

Cauliflower Dip

Preparation Time: 10 minutes

Cooking Time: 40 minutes

Servings: 10

Ingredients:

1 cauliflower head, cut into florets

1 1/2 cups parmesan cheese, shredded

2 tbsp green onions, chopped

2 garlic clove

1 tsp Worcestershire sauce

1/2 cup sour cream

3/4 cup mayonnaise

8 oz cream cheese, softened

2 tbsp olive oil

Directions:

Toss cauliflower florets with olive oil.

Add cauliflower florets into the air fryer basket and cook at 390 F for 20-25 minutes. Shake basket halfway through.

Add cooked cauliflower, 1 cup parmesan cheese, green onion, garlic, Worcestershire sauce, sour cream, mayonnaise, and cream cheese into the food processor and process until smooth.

Transfer cauliflower mixture into the 7-inch dish and top with remaining parmesan cheese.

Place dish in air fryer basket and cook at 360 F for 10-15 minutes.

Serve and enjoy.

Nutrition:

Calories 308

Fat 29 g

Carbohydrates 3 g

Sugar 1 g

Protein 7 g

Cholesterol 51 m

VEGETABLES AND SIDE RECIPES

Cajun Style French Fries

Preparation time: 30 minutes

Cooking time: 28 minutes

Servings: 4

Ingredients:

2 reddish potatoes, peeled and cut into strips of 76 x 25 mm

1 liter of cold water

15 ml of oil

7g of Cajun seasoning

1g cayenne pepper

Tomato sauce or ranch sauce, to serve

Direction:

Cut the potatoes into 76 x 25 mm strips and soak them in water for 15 minutes.

Drain the potatoes, rinse with cold, dry water with paper towels.

Preheat the air fryer, set it to 195°C.

Add oil and spices to the potatoes, until they are completely covered.

Add the potatoes to the preheated air fryer and set the timer to 28 minutes.

Be sure to shake the baskets in the middle of cooking

Remove the baskets from the air fryer when you have finished cooking and season the fries with salt and pepper.

Serve with tomato sauce or ranch sauce.

Nutrition:

Calories: 156

Fat: 8.01g

Carbohydrate: 20.33g

Protein: 1.98g

Sugar: 0.33g

Cholesterol: 0mg

Vegetables In Air Fryer

Preparation time: 20 minutes

Cooking time: 30 minutes

Servings: 2

Ingredients:

2 potatoes

1 zucchini

1 onion

1 red pepper

1 green pepper

Direction:

Cut the potatoes into slices.

Cut the onion into rings.

Cut the zucchini slices

Cut the peppers into strips.

Put all the ingredients in the bowl and add a little salt, ground pepper and some extra virgin olive oil.

Mix well.

Pass to the basket of the Air fryer.

Select 1600C, 30 minutes.

Check that the vegetables are to your liking.

Nutrition:

Calories: 135

Fat: 11g

Carbohydrates: 8g

Protein: 1g

Sugar: 2g

Cholesterol: 0mg

Crispy Rye Bread Snacks With Guacamole And Anchovies

Preparation time: 10 minutes

Cooking time: 10 minutes

Servings: 4

Ingredients:

4 slices of rye bread

Guacamole

Anchovies in oil

Direction

Cut each slice of bread into 3 strips of bread.

Place in the basket of the Air fryer, without piling up, and we go in batches giving it the touch you want to give it. You can select 1800C, 10 minutes.

When you have all the crusty rye bread strips, put a layer of guacamole on top, whether homemade or commercial.

In each bread, place 2 anchovies on the guacamole.

Nutrition:

Calories: 180

Fat: 11.6g

Carbohydrates: 16g

Protein: 6.2g

Sugar: 0g

Cholesterol: 19.6mg

Mushrooms Stuffed With Tomato

Preparation time: 5 minutes

Cooking time: 50 minutes

Servings: 4

Ingredients:

8 large mushrooms

250g of minced meat

4 cloves of garlic

Extra virgin olive oil

Salt

Ground pepper

Flour, beaten egg and breadcrumbs

Frying oil

Fried Tomato Sauce

Direction:

Remove the stem from the mushrooms and chop it. Peel the garlic and chop. Put some extra virgin olive oil in a pan and add the garlic and mushroom stems.

Sauté and add the minced meat. Sauté well until the meat is well cooked and season.

Fill the mushrooms with the minced meat.

Press well and take the freezer for 30 minutes.

Pass the mushrooms with flour, beaten egg and breadcrumbs.

Beaten egg and breadcrumbs.

Place the mushrooms in the basket of the Air fryer.

Select 20 minutes, 1800C.

Distribute the mushrooms once cooked in the dishes.

Heat the tomato sauce and cover the stuffed mushrooms.

Nutrition:

Calories: 160

Fat: 7.96g

Carbohydrates: 19.41g

Protein: 7.94g

Sugar: 9.19g

Cholesterol: 0mg

Spiced Potato Wedges

Preparation time: 15

Cooking time: 40 minutes

Servings: 4

Ingredients:

8 medium potatoes

Salt

Ground pepper

Garlic powder

Aromatic herbs, the one we like the most

2 tbsp extra virgin olive oil

4 tbsp breadcrumbs or chickpea flour

Direction:

Put the unpeeled potatoes in a pot with boiling water and a little salt.

Let cook 5 minutes. Drain and let cool. Cut into thick segments, without peeling.

Put the potatoes in a bowl and add salt, pepper, garlic powder, the aromatic herb that we have chosen oil and breadcrumbs or chickpea flour.

Stir well and leave 15 minutes. Pass to the basket of the Air fryer and select 20 minutes, 1800C.

From time to time shake the basket so that the potatoes mix and change position. Check that they are tender.

Nutrition:

Calories: 121

Fat: 3g

Carbohydrates: 19g

Protein: 2g

Sugar: 0g

Cholesterol: 0mg

Egg Stuffed Zucchini Balls

Preparation time: 15 minutes

Cooking time: 45-60 minutes

Servings: 4

Ingredients:

2 zucchini

1 onion

1 egg

120g of grated cheese

4 eggs

Salt

Ground pepper

Flour

Direction:

Chop the zucchini and onion in the Thermomix, 10 seconds speed 8, in the Cuisine with the kneader chopper at speed 10 about 15 seconds or we can chop the onion by hand and the zucchini grate. No matter how you do it, the important thing is that the zucchini and onion are as small as possible.

Put in a bowl and add the cheese and the egg. Pepper and bind well.

Incorporate the flour, until you have a very brown dough with which you can wrap the eggs without problems.

Cook the eggs and peel.

Cover the eggs with the zucchini dough and pass through the flour.

Place the four balls in the basket of the Air fryer and paint with oil.

Select 1800C and leave for 45 to 60 minutes or until you see that the balls are crispy on the outside.

Serve over a layer of mayonnaise or aioli.

Nutrition:

Calories: 23

Fat: 0.5g

Carbohydrates: 2g

Protein: 1.8g

Sugar: 0g

Cholesterol: 15mg

Vegetables With Provolone

Preparation time: 10 minutes

Cooking time: 30 minutes

Servings: 4

Ingredients:

1 bag of 400g of frozen tempura vegetables

Extra virgin olive oil

Salt

1 slice of provolone cheese

Direction:

Put the vegetables in the basket of the Air fryer. Add some strands of extra virgin olive oil and close.

Select 20 minutes, 2000C.

Pass the vegetables to a clay pot and place the provolone cheese on top.

Take to the oven, 1800C, about 10 minutes or so or until you see that the cheese has melted to your liking.

Nutrition:

Calories: 104

Fat: 8g

Carbohydrates: 0g

Protein: 8g

Sugar: 0g

Cholesterol: 0mg

FISH AND SEAFOOD RECIPES

Tasty Air Fried Cod

Preparation Time: 22 minutes

Servings: 4

Ingredients

7 oz. 2 cod fish

Sesame oil

Salt and black pepper

1 cup water

1 tsp. dark soy sauce

4 tbsp. light soy sauce

1 tbsp. sugar

3 tbsp. olive oil

4 ginger slices

3 spring onions

2 tbsp. coriander

Direction

Season fish with pepper, salt, sprinkle sesame oil, rub well and allow for 10 minutes.

Add fish to air fryer. Cook at 356°F for 12 minutes.

Heat a pot with the water over medium heat. Add sugar and light and dark soy sauce. Allow to simmer. Take off heat.

Heat pan with the olive oil over medium heat. Add green onions and ginger. Cook for a few minutes. Take off heat.

Divide fish on plates. Top with ginger and green onions. Drizzle soy sauce mix. Sprinkle coriander and serve.

Delicious Catfish

Preparation Time: 30 minutes

Servings: 4

Ingredients

4 cat fish fillets

Black pepper and Salt

A pinch of sweet paprika

1 tbsp. parsley

1 tbsp. lemon juice

1 tbsp. olive oil

Direction

Season catfish fillets with salt, paprika, pepper, drizzle oil, rub well. Then put in air fryer basket and cook at 400°F for 20 minutes. Flip the fish after 10 minutes of time.

Share fish on plates. Sprinkle parsley and drizzle some lemon juice over it, serve.

Cod Fillets with Fennel and Grapes Salad

Time: 25 minutes

Servings: 2

Ingredients

2 black cod fillets

1 tbsp. olive oil

Black pepper and Salt

1 fennel bulb

1 cup grapes

½ cup pecans

Directions

Sprinkle half of the oil over fish fillets, season with pepper and salt, rub well, place fillets in air fryer basket. Then cook for 10 minutes' time at 400°F and put in plate.

Mix pecans with grapes, fennel, the rest of the oil, salt and pepper, toss to coat, in a bowl. Add to pan that fits air fryer. Cook at 400° F for 5 minutes.

Share cod on plates, add grapes and fennel mix on the side then serve.

Tabasco Shrimp

Preparation Time: 20 minutes

Servings: 4

Ingredients

1 lb. shrimp

1 tbsp. red pepper flakes

2 tbsp. olive oil

1 tbsp. Tabasco sauce

2 tbsp. water

1 tbsp. oregano

black pepper and salt

½ tbsp. parsley

½ tbsp. smoked paprika

Direction

Mix oil with water, pepper flakes, Tabasco sauce, oregano, parsley, pepper, salt, paprika and shrimp and toss well to coat in a bowl

Place shrimp to preheated air fryer at 370° F and cook for 10 minutes. Shake fryer once.

Share shrimp on plates then serve by a side salad.

Buttered Shrimp Skewers

Time: 16 minutes

Servings: 2

Ingredients

8 shrimps

4 garlic cloves

Black pepper and Salt

8 green bell pepper

1 tbsp. rosemary

1 tbsp. butter

Direction

Mix shrimp with garlic, pepper, salt, butter, bell pepper slices and rosemary, toss to coat in a bowl, and allow for 10 minutes.

Place 2 bell pepper slices and 2 shrimp on a skewer. Repeat for the remaining shrimp and bell pepper pieces.

Place them all in your air fryer basket and cook at 360°F for 6 minutes.

Share on plates and serve.

Asian Salmon

Preparation Time: 75 minutes

Servings: 2

Ingredients

2 medium salmon fillets

6 tbsp. light soy sauce

3 tbsp. mirin

1 tbsp. water

6 tbsp. honey

Directions

Mix soy sauce with honey, water and mirin, whisk very well, in a bowl, add salmon, rub well. Leave it in a fridge for an hour.

Place salmon to air fryer. Cook at 360° F for 15 minutes' time. Flip it after 7 minutes.

Put the soy marinade in pan, heat above medium heat, whisk very well. Cook for 2 minutes the put off heat.

Share salmon on plates, sprinkle marinade over and serve.

Cod Steaks with Plum Sauce

Preparation Time: 20 minutes

Servings: 2

Ingredients

2 big cod steaks

Salt and black pepper

½ tbsp. garlic powder

½ tbsp. ginger powder

¼ tbsp. turmeric powder

1 tbsp. plum sauce

Cooking spray

Direction

Season cod steaks with pepper and salt. Spray with cooking oil and add ginger powder, turmeric powder and garlic powder then rub well.

Place the cod steaks in air fryer. Cook at 360°F for 15 minutes, after 7 minutes flip them.

Heat pan above medium heating. Add plum sauce and stir. Cook for 2 minutes.

Share cod steaks on plates, sprinkle plum sauce over and serve

POULTRY RECIPES

Country Chicken Tenders

Preparation Time: 30 minutes

Servings: 3

Ingredients:

3/4 pound chicken tenders

For the breading

1/2 cup seasoned breadcrumbs

2 tablespoons olive oil

2 eggs (beaten)

1 teaspoon black pepper

1/2 teaspoon salt

1/2 cup all-purpose flour

Directions:

Combine breadcrumbs and salt in a bowl. Add olive oil and mix well.

Put the beaten eggs in a different bowl and flour in another.

Toss meat in the bowl of flour until coated. Dip them in egg and coat with the breadcrumb mixture. Press using your hands to make the coating stick to the meat. Arrange them in the cooking basket.

Cook for 10 minutes at 330 degrees. Turn the temperature to 350 degrees and cook for 5 more minutes.

Nutrition: Calories: 486 calories, 21.6g fat, 29.5g carbohydrates, 41g protein

Buffalo Chicken Tenders

Preparation Time: 34 minutes

Servings: 4

Ingredients:

1/2 cup Buffalo sauce

1 cup flour

1 pound chicken tenders (trimmed)

1 cup ranch dressing

1/4 cup blue cheese (crumbled)

1/2 teaspoon garlic powder

1/2 teaspoon cayenne pepper

1/2 teaspoon salt

Directions:

Put the ranch dressing in a bowl. Add the meat and leave to marinate for an hour.

Mix to combine flour, cayenne pepper, salt, and garlic powder in a bowl. Dip each chicken piece in the mixture until coated. Put 2 chicken tenders in the cooking basket at a time

Cook for 13 minutes at 400 degrees. Shake the basket twice during the cooking process. Transfer the cooked meat to a bowl and cook the rest.

Put the buffalo sauce in a bowl. Add the cooked meat and toss until coated. Transfer to a plate and sprinkle with cheese before serving.

Nutrition: Calories: 380 calories, 12.1g fat, 28g carbohydrates, 41g protein

Korean BBQ Satay

Preparation Time: 22 minutes

Servings: 3

Ingredients:

1 tablespoon grated fresh ginger

2 teaspoons toasted sesame seeds

1/2 cup pineapple juice

1/2 cup low-sodium soy sauce

1/4 cup sesame oil

4 garlic cloves (chopped)

12 ounces chicken tenders (boneless and skinless)

4 scallions (chopped)

A pinch of black pepper

Directions:

Skewer each piece of meat and trim excess fat.

Put the remaining ingredients in a bowl. Mix well. Add the skewered chicken and make sure that all pieces are covered with the mixture. Cover the bowl and refrigerate for 2 hours or overnight.

Use paper towels to pat the meat dry. Arrange the skewers in the cooking basket. Cook for 7 minutes at 390 degrees.

Nutrition: Calories: 442 calories, 27.8g fat, 12.6g carbohydrates, 41g protein

Jerk Chicken Wings

Preparation Time: 33 minutes

Servings: 5

Ingredients:

5 tablespoons lime juice

1 habanero pepper (remove the ribs and seeds, chopped)

1/2 cup red wine vinegar

3 pounds chicken wings

6 garlic cloves (minced)

2 tablespoons soy sauce

2 tablespoons olive oil

1 teaspoon white pepper

1 teaspoon salt

1 teaspoon cinnamon

1 teaspoon cayenne pepper

1 tablespoon grated fresh ginger

1 tablespoon chopped fresh thyme

1 tablespoon allspice

2 tablespoons brown sugar

4 scallions (minced)

Directions:

Put all ingredients in a bowl and mix until combined and the meat is well-coated. Transfer to a Ziploc bag. Refrigerate for 2 hours or overnight.

Drain the liquid and pat the meat dry using paper towels. Arrange them in the cooking basket.

Cook for 18 minutes at 390 degrees. Shake the basket halfway through the cooking process.

Note: Serve the dish along with your favorite sauces. You can also try it with ranch dressing or blue cheese dipping sauce.

Nutrition: Calories: 623 calories, 27.8g fat, 13.6g carbohydrates, 801g protein

Fried Chicken Tenders with Mustard and Sage

Preparation Time: 30 minutes

Servings: 2

Ingredients:

1 tablespoon melted butter

1 tablespoon mayonnaise

1/2 cup panko breadcrumbs

1/2 teaspoon dry sage

4 chicken tenders

1 teaspoon Dijon mustard

Directions:

Mix to combine sage, mustard, and mayonnaise in a bowl.

In another bowl, put the butter and breadcrumbs and mix well.

Pat the meat dry using paper towels. Coat each piece with a bit of the mayonnaise mixture and cover with the breadcrumb mixture.

Arrange in a single layer in the cooking basket. Cook for 10 minutes at 392 degrees. Flip the meat and cook for 10 more minutes.

Nutrition: Calories: 727 calories, 30.9g fat, 21.6g carbohydrates, 85g protein

Parmesan Crusted Chicken Fillet

Preparation Time: 20 minutes

Servings: 4

Ingredients:

1 teaspoon Italian herbs

1 teaspoon garlic powder

1 egg

30 grams salted butter (melted)

1/2 cup parmesan cheese

1 cup panko breadcrumbs

8 pieces chicken tenders or fillet

Directions:

In a bowl, mix to combine melted butter, egg, Italian herbs, and garlic powder. Add the meat and marinate for at least an hour.

Put the Parmesan cheese and panko breadcrumbs in a shallow bowl. Mix until combined.

Drain liquid from the meat and coat each piece with the breadcrumb and cheese mixture. Leave for 5 minutes.

Line the Air Fryer's base with aluminum foil. Arrange 4 pieces of the coated meat in the basket for each batch. Cook for 6 minutes at 200 degrees. Transfer to a platter and cook the rest of the meat.

Serve while hot.

Nutrition: Calories: 669 calories, 24g fat, 21.6g carbohydrates, 85g protein

BEEF AND LAMB RECIPES

Mediterranean Lamb Meatballs

Preparation time: 10 minutes

Cooking time: 40 minutes

Servings: 4

Ingredients:

454g ground lamb

3 cloves garlic, minced

5g of salt

1g black pepper

2g of mint, freshly chopped

2g ground cumin

3 ml hot sauce

1g chili powder

1 scallion, chopped

8g parsley, finely chopped

15 ml of fresh lemon juice

2g lemon zest

10 ml of olive oil

Direction:

Mix the garlic, lamb, salt, cumin, pepper, mint, hot sauce, chili powder, cumin, chives, lemon juice, parsley, and lemon zest until well combined.

Create balls with the lamb mixture and cool for 30 minutes.

Select Preheat in the air fryer and press Start/Pause.

Cover the meatballs with olive oil and place them in the preheated fryer.

Select Steak, set the time to 10 minutes and press Start/Pause.

Nutritional Value (Nutrition per Serving):

Calories: 282

Fat: 23.41

Carbohydrates: 0g

Protein: 16.59

Sugar: 0g

Cholesterol: 73gm

North Carolina Style Pork Chops

Preparation time: 5 minutes

Cooking time: 10 minutes

Servings: 2

Ingredients:

2 boneless pork chops

15 ml of vegetable oil

25g dark brown sugar, packaged

6g of Hungarian paprika

2g ground mustard

2g freshly ground black pepper

3g onion powder

3g garlic powder

Salt and pepper to taste

Direction:

Preheat the air fryer a few minutes at 180oC.

Cover the pork chops with oil.

Put all the spices and season the pork chops abundantly, almost as if you were making them breaded.

Place the pork chops in the preheated air fryer.

Select Steak, set the time to 10 minutes.

Remove the pork chops when it has finished cooking. Let it stand for 5 minutes and serve.

Nutrition:

Calories: 118

Fat: 6.85g

Carbohydrates: 0

Protein: 13.12g

Sugar: 0g

Cholesterol: 39mg

Beef With Sesame And Ginger

Preparation time: 10 minutes

Cooking time: 23 minutes

Servings: 4-6

Ingredients:

½ cup tamari or soy sauce

3 tbsp olive oil

2 tbsp toasted sesame oil

1 tbsp brown sugar

1 tbsp ground fresh ginger

3 cloves garlic, minced

1 to 1½ pounds skirt steak, boneless sirloin or low loin

Direction:

Put together the tamari sauce, oils, brown sugar, ginger and garlic in small bowl. Add beef to a quarter-size plastic bag and pour the marinade into the bag. Press on the bag as much air as possible and seal it.

Refrigerate for 1 to 1½ hours, turning half the time. Remove the meat from the marinade and discard the marinade. Dry the meat with paper towels. Cook at a temperature of 350°F for 20 to 23 minutes, turning halfway through cooking.

Nutrition:

Calories: 381

Fat: 5g

Carbohydrates: 9.6g

Protein: 38g

Sugar: 1.8g

Cholesterol: 0mg

Provencal Ribs

Preparation time: 10 minutes

Cooking time: 1h 20 minutes

Servings: 4

Ingredients:

500g of pork ribs

Provencal herbs

Salt

Ground pepper

Oil

Direction:

Put the ribs in a bowl and add some oil, Provencal herbs, salt and ground pepper.

Stir well and leave in the fridge for at least 1 hour.

Put the ribs in the basket of the Air fryer and select 2000C, 20 minutes.

From time to time shake the basket and remove the ribs.

Nutrition:

Calories: 296

Fat: 22.63g

Carbohydrates: 0g

Protein: 21.71g

Sugar: 0g

Cholesterol: 90mg

Beef Scallops

Preparation time: 15 minutes

Cooking time: 20 minutes

Servings: 4

Ingredients:

16 veal scallops

Salt

Ground pepper

Garlic powder

2 eggs

Bread crumbs

Extra virgin olive oil

Direction:

Put the beef scallops well spread and salt and pepper. Add some garlic powder.

In a bowl, beat the eggs.

In another bowl put the breadcrumbs.

Pass the Beef scallops for beaten egg and then for the breadcrumbs.

Spray with extra virgin olive oil on both sides.

Put a batch in the basket of the Air fryer. Do not pile the scallops too much.

Select 1800C, 15 minutes. From time to time shake the basket so that the scallops move.

When finishing that batch, put the next one and so on until you finish with everyone, usually 4 or 5 scallops enter per batch.

Nutrition:

Calories: 330

Fat: 16.27g

Carbohydrates: 0g

Protein: 43g

Sugar: 0g

Cholesterol: 163mg

Potatoes With Loin And Cheese

Preparation time: 10 minutes

Cooking time: 30 minutes

Servings: 4

Ingredients:

1kg of potatoes

1 large onion

1 piece of roasted loin

Extra virgin olive oil

Salt

Ground pepper

Grated cheese

Direction:

Peel the potatoes, cut the cane, wash and dry.

Put salt and add some threads of oil, we bind well.

Pass the potatoes to the basket of the Air fryer and select 1800C, 20 minutes.

Meanwhile, in a pan, put some extra virgin olive oil and add the peeled onion and cut into julienne.

When the onion is transparent, add the chopped loin.

Saute well and pepper.

Put the potatoes on a baking sheet.

Add the onion with the loin.

Cover with a layer of grated cheese.

Bake a little until the cheese takes heat and melts.

Nutrition:

Calories: 332

Fat: 7g

Carbohydrates: 41g

Protein: 23g

Sugar: 0g

Cholesterol: 0mg

PORK RECIPES

Tortilla Crusted Pork Loin Chops

Preparation Time: 30 minutes

Servings: 2

Ingredients:

1 teaspoon sauce (hot sauce, steak sauce or Worcestershire sauce)

1/2 teaspoon salt

2 pork loin chops (boneless)

1 egg (beaten)

1/2 cup buttermilk

1/2 cup crushed tortilla chips

1/2 cup flour

Directions:

Put salt, Worcestershire sauce, and buttermilk in a Ziploc bag. Seal and shake until combined. Add the meat and shake until all sides are coated. Refrigerate overnight to marinate.

Drain liquid from the meat and use paper towels to pat excess liquid.

Put the beaten egg in a bowl, crushed chips in another bowl, and the flour in a different bowl.

Coat the meat with flour, dip in the egg and press into the crushed chips until all sides are covered. Arrange them in the cooking basket.

Cook for 15 minutes at 356 degrees. Flip the meat halfway through the cooking process.

Transfer to a plate and leave to rest for 5 minutes before serving.

Nutrition: Calories: 463 calories, 27g fat, 32g carbohydrates, 25.1g protein

Nutrition: Calories: 962 calories, 68.6g fat, 32g carbohydrates, 53.1g protein

Jalapeno Bacon Poppers

Preparation Time: 40 minutes

Servings: 4

Ingredients:

12 bacon slices (thin)

1/4 teaspoon black pepper

1/4 teaspoon sea salt

3 green onions (minced)

1/2 cup sharp white cheddar cheese (freshly shredded)

2 tablespoons Greek Yogurt

12 jalapeno peppers (large)

Directions:

Place each pepper in a chopping board and cut into 2 in a lengthwise manner. Remove the membranes and seeds.

Put the black pepper, salt, green onion, cheddar cheese, and yogurt in a blender. Process until smooth. Transfer to a bowl, cover, and refrigerate until firm.

Scoop a small portion of the filling in the half part of a pepper and place the top part on top to enclose the filling. Wrap each filled pepper with a slice of bacon and secure it with a toothpick.

Arrange them in the cooking basket. Cook for 15 minutes at 400 degrees.

Nutrition: Calories: 463 calories, 31.6g fat, 9.1g carbohydrates, 35g protein

Jamaican Jerk Pork

Preparation Time: 30 minutes

Servings: 4

Ingredients:

1/4 cup jerk paste

1.5 pounds pork butt (chopped into 3 chunks)

Directions:

Rub meat with the jerk paste. Refrigerate for at least 4 hours or overnight to marinate. Leave at a room temperature for 20 minutes before cooking.

Nutrition: Calories: 338 calories, 12g fat, 1.1g carbohydrates, 53g protein

Roasted Pork

Preparation time: 5 minutes

Cooking time: 30 minutes

Servings: 2-4

Ingredients:

500-2000g Pork meat (To roast)

Salt

Oil

Direction:

Join the cuts in an orderly manner.

Place the meat on the plate

Varnish with a little oil.

Place the roasts with the fat side down.

Cook in air fryer at 1800C for 30 minutes.

Turn when you hear the beep.

Remove from the oven. Drain excess juice.

Let stand for 10 minutes on aluminum foil before serving.

Nutrition:

Calories: 820

Fat: 24.75g

Carbohydrates: 117g

Protein: 33.75g

Sugar: 0g

Cholesterol: 120mg

DESSERTS

Sweet Sponge Cake

Preparation time: 15 minutes

Cooking time: 60 minutes

Servings: 10

Ingredients:

250g baking powder with yeast

250g of sugar

3 medium eggs

3 tbsp olive oil Grated orange

300g chopped pistachio

1 sachet of yeast

Lemon cream:

1 egg white

150g luster sugar

100 ml sour cream

1 tsp lemon juice

Direction:

Separate the yolks from the eggs. Mount the egg whites until stiff with the blender and gradually incorporate the sugar. Mix until you get a thick white cream. Apart, beat the yolks with the oil and orange zest.

Incorporate this mixture with the clear ones, mix in an enveloping way and finally incorporate the flour and the yeast with a sieve. When everything is well mixed, add the pistachios. You can use a circular mold greased with oil and flour or kitchen paper that is more comfortable.

Add the cake dough to the mold. Preheat the Air fryer a few minutes to 1600C. Insert the mold into the basket of the Air fryer and set the timer for about 30 minutes at 160°C. While it is cooked prepare the lemon cream.

To do so, gradually mix the white with the sugar, add the lemon juice and add Sour cream and mix until you get a thick cream. Serve the sponge cake with the lemon cream on top and sprinkle with chopped pistachios.

Nutrition:

Calories: 495

Fat: 23g

Carbohydrates: 62g

Protein: 10g

Sugar: 300g

Cholesterol: 67mg

Egg Flan

Preparation time: 15 minutes

Cooking time: 60 minutes

Servings: 4

Ingredients:

300 ml of milk

3 eggs

80g of sugar

Direction:

Put the sugar in a saucepan reserving two tablespoons for later. Add some water. With very low heat melt the sugar until everything is liquid and caramelized.

Immediately pour into the molds for custards (whether they are individual or if it is a large flan). It is important to do it right away because the caramel solidifies very quickly when it cools.

In a separate bowl, beat the eggs with the help of some rods. When they begin to foam, add the milk and mix everything very well.

Once the mixture is homogeneous pour into the molds to which we have previously put the candy.

Next, preheat the Air fryer a few minutes to 1600C. Then cook the custards in a water bath in the Ai fryer. To do so, arrange the flan inside the basket of the Air fryer in a container with water ensuring that water reaches half of the containers but ensuring that no water enters them.

Put the bowl with the custards and with the water half by bathing them in the Air fryer and let everything cook at medium temperature 160 C for about 1 hour.

To check if the custards are cooked, shake gently and if, when moving, they have the consistent appearance of the custards, they are ready. Otherwise, if they look very liquid, bake them in the water bath a little more.

Nutrition:

Calories: 175

Fat: 6g

Carbohydrates: 24g

Protein: 7g

Sugar: 80g

Cholesterol: 160mg

Roasted Apples

Preparation time: 10 minutes

Cooking time: 20 minutes

Servings: 4

Ingredients:

4 apples

4 tsp butter

4 tsp honey

A little cinnamon powder

Direction:

Separate apples to remove the shape of the heart.

Incorporate, in the center of each apple, a teaspoon of butter, another of honey and a little cinnamon.

Preheat the Air fryer a few minutes at 1800C.

Put the apples in the basket of the Air fryer and set the timer 20 minutes at 1800C

Nutrition:

Calories: 179

Fat: 11g

Carbohydrates: 20g

Protein: 0g

Sugar: 50g

Cholesterol: 31mg

Homemade Muffins

Preparation time: 10 minutes

Cooking time: 15 minutes

Servings: 3

Ingredients:

6 tbsp olive oil

100g of sugar

2 eggs

100g flour

1 tsp Royal baking powder

Lemon zest

Direction:

Beat the eggs with the sugar, with the help of a whisk. Add the oil little by little, while stirring, until you get a fluffy cream. Then add the lemon zest.

Finally, add the sifted flour with the yeast to the previous mixture and mix in an envelope.

Fill 2/3 of the muffin muffins with the dough.

Preheat the Air fryer a few minutes to 1800C and when ready place the muffins in the basket.

Set the timer for approximately 20 minutes at a temperature of 1800C, until they are golden brown.

Nutrition:

Calories: 240

Fat: 12g

Carbohydrates: 29g

Protein: 4g

Sugar: 100g

Cholesterol: 67g

Palm Trees Hojaldre

Preparation time: 5 minutes

Cooking time: 15 minutes

Servings: 2

Ingredients:

1 Sheet of puff pastry

Sugar

Direction:

Stretch the puff pastry sheet.

Pour the sugar over and fold the puff pastry sheet in half.

Put a thin layer of sugar on top and fold the puff pastry in half again.

Roll the puff pastry sheet from both ends towards the center (creating the shape of the palm tree).

Cut into sheets 5-8 mm thick.

Preheat the Air fryer to 1800C and put the palm trees in the basket.

Set the timer about 10 minutes at 1800C.

Nutrition:

Calories: 317

Fat: 17g

Carbohydrates: 38g

Protein: 3g

Sugar: 40g

Cholesterol: 23g

Chocolate And Nut Cake

Preparation time: 10 minutes

Cooking time: 30 minutes

Servings: 4

Ingredients:

60g dark chocolate

2 butter spoons

1 egg

3 spoonfuls of sugar

50g flour

1 envelope Royal yeast

Chopped walnuts

Direction:

Melt the dark chocolate with the butter, over low heat. Once melted, put in a bowl.

Incorporate the egg, sugar, flour, yeast (the latter passed through the sieve, to prevent lumps from formingand finally the chopped nuts.

Beat well by hand until you get an uniform dough.

Put the dough in a silicone mold or oven suitable for incorporation in the basket of the Air fryer.

Preheat the Air fryer a few minutes at 1800C.

Set the timer for 20 minutes at 1800C and when it has cooled down, unmold.

Nutrition:

Calories: 108

Fat: 4g

Carbohydrates: 16g

Protein: 2g

Sugar: 250g

Cholesterol: 3mg

Light Cheese Cake With Strawberry Syrup

Preparation time: 10 minutes

Cooking time: 20 minutes

Servings: 4

Ingredients:

500g cottage cheese

3 whole eggs

2 tbsp powdered sweetener

2 tbsp oat bran

½ tbsp baking yeast

2 tbsp cinnamon

2 tbsp vanilla aroma

1 lemon (the skin

Direction:

Mix in a bowl the cottage cheese, the sweetener, the cinnamon, the vanilla aroma and the lemon zest. Mix very well until you get a homogeneous cream.

Incorporate the eggs one by one.

Finally, add oats and yeast mixing well.

Put the whole mixture in a container to fit in the Air fryer.

Preheat the Air fryer a few minutes at 1800C.

Insert the mold into the basket of the Air fryer and set the timer for about 20 minutes at 180°C.

Nutrition:

Calories: 191

Fat: 9g

Carbohydrates: 21g

Protein: 4g

Sugar: 18g

Cholesterol: 0mg